AF715098

Because of a Storm: The Spongy Moth: A Small Bug with a Big (and Icky) Impact
Written by Dorna Schroeter
Illustrated by Nataliya Chervochkova

Publication may not be reproduced, distributed, or transmitted in any form or by any means, including photocopying, recording, or other electronic or mechanical methods without the prior written permission of the publisher, except in cases of brief quotations embodied in critical reviews and certain other noncommercial uses permitted by copyright law. For permission request, contact inspiredbynaturebooks.com.

Inspired by Nature Books
Rhinebeck, NY 12572
For teaching resources and more, visit inspiredbynaturebooks.com

Publisher's Cataloging-in-Publication Data
Names: Schroeter, Dorna, author | Chervochkova, Nataliya, illustrator.

Title: Because of a Storm: The Spongy Moth: A Small Bug with a Big (and Icky) Impact
Written by Dorna Schroeter and illustrated by Nataliya Chervochkova
Description: First Edition | Rhinebeck, NY: Inspired by Nature Books, 2025
Audience: Ages 8-11

Summary: Raining poop! Clouds of moths! Leafless summer trees! Welcome to a spongy moth outbreak! This bug-filled adventure tells the story of an invasive species that's been bugging North America for over 150 years. Travel back to the 1800s to learn how this problem started. Meet a modern-day scientist who has been studying these pesky bugs for over 30 years.

LCCN: 2025914667 | ISBN | 978-1-966997-01-6 (hardcover) | 978-1-966997-00-9 (paperback) | 978-1-966997-02-3 (ebook)

Subjects: LCSH: Science & Nature / Environmental Conservation & Protection--JUVENILE NONFICTION. | Science & Nature / Environmental Science & Ecosystems--JUVENILE NONFICTION. | History /United States / 19th Century--JUVENILE NONFICTION. | BISAC: JUVENILE NONFICTION /Science & Nature / Environmental Conservation & Protection / JUVENILE NONFICTION / Science & Nature / Environmental Science & Ecosystems / JUVENILE NONFICTION / History / United States /19th Century

Classification: LCC PZ7.1.S336548 Be 2025 | DDC [E] - dc23
Lexile: 780L

Because of a Storm

The Spongy Moth:
A Small Bug with a Big (and Icky) Impact

By Dorna Schroeter

Illustrated by
Natalia Chervochkova

Sometimes, humans try to solve one problem and end up creating new ones.

This is a story about one of those problems.

It begins in America in the mid-1800s with . . .

SILK!

Silk is a fabric made from the silk strands produced by silkworms. They make these strands when spinning their cocoons.

For thousands of years, people have highly valued silk fabric. It has an ***iridescent*** shine, is soft as kitten fur and strong as steel.

Making silk requires

LOTS of Silkworms

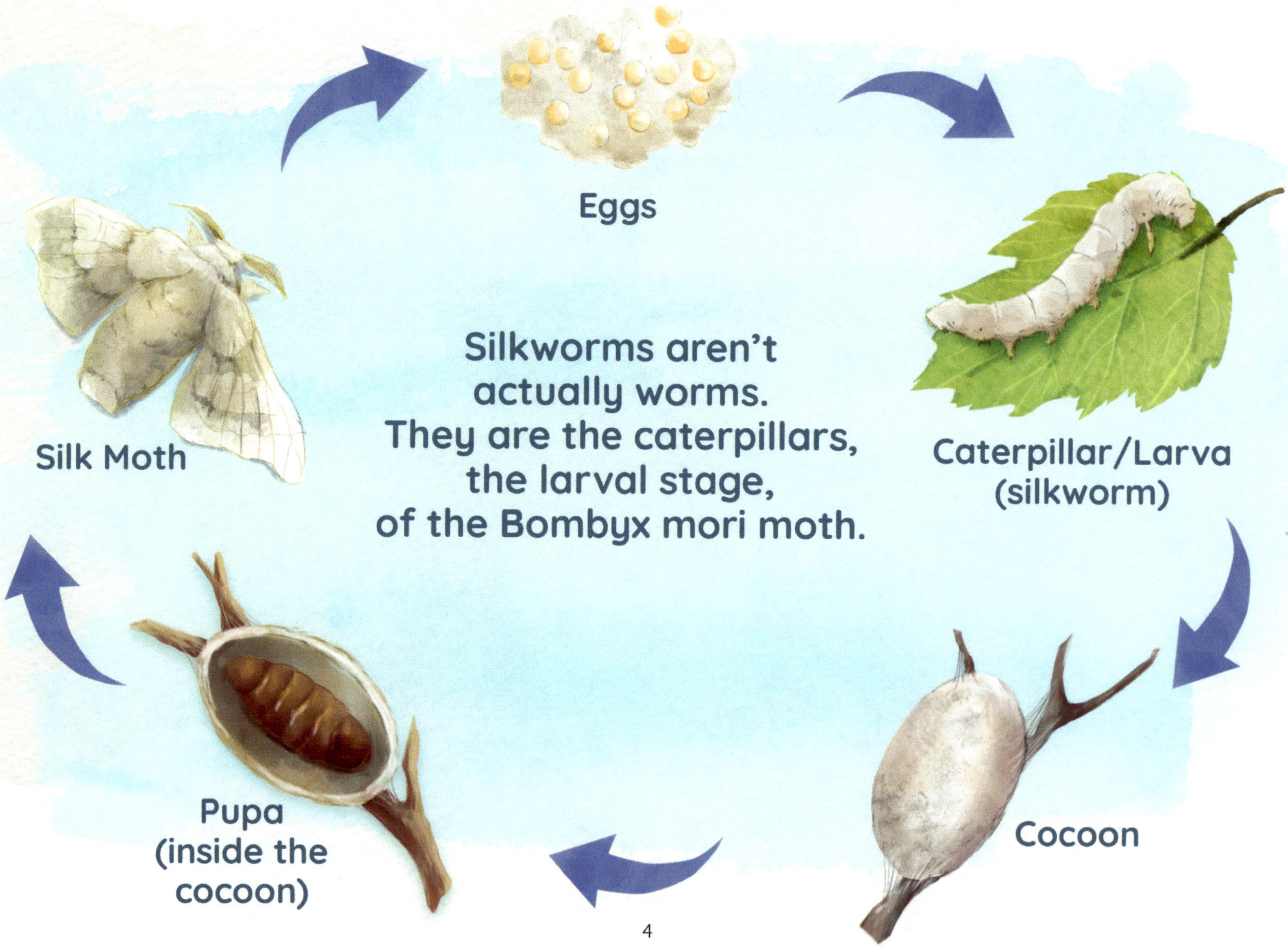

Silkworms aren't actually worms. They are the caterpillars, the larval stage, of the Bombyx mori moth.

LOTS of mulberry tree leaves
(the only food they eat)

People

A protected space

And LOTS of time

In the early 1800s, people in New England began making silk at home. It was one way for them to earn a living. Families, church groups, and whole villages began making silk thread.

Men planted and cared for the mulberry trees.

Children cared for silkworms.

They fed them freshly picked leaves throughout the day. To prevent disease, they removed the caterpillars' droppings and shed ***exoskeletons***.

The leaves must be fresh, as silkworms don't drink water. The leaves supply their water needs.

Some families raised them inside their homes. Others built a cocoonery.

Women and girls unraveled each silkworm cocoon.

They then spun the strands together to produce silk thread. The thread was sold to textile mills. There, machines wove it into silk fabric.

Each silkworm produces just one single strand of silk, which measures 2,000 - 3,000 feet long and is held together by a type of natural gum, called sericin.

Some families produced as much as 500 pounds of silk thread a year!

But then...

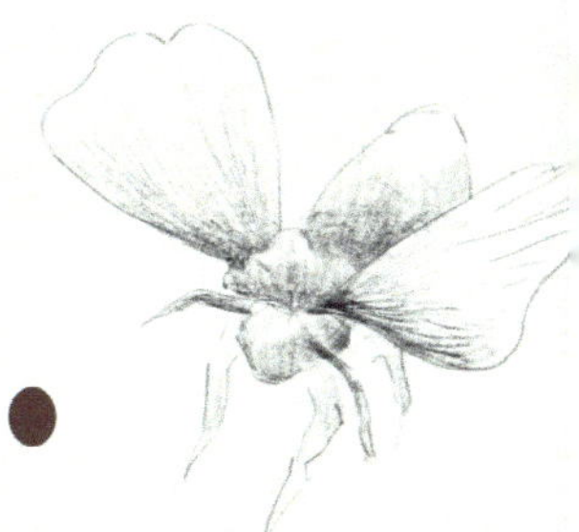

Several diseases began killing silkworms.

Then, harsh winter weather and a ***plague*** began killing mulberry trees.

How could people continue to make a living this way without silkworms?

Étienne Trouvelot

was an amateur ***entomologist***
living in Massachusetts.

Hearing about these problems, he wondered if he could help.

Could he produce a hardier silkworm by crossbreeding it with another silk-producing caterpillar?

From his research, he learned about

Spongy Moths.

In their caterpillar stage, they produce silk strands to use in moving from tree to tree. Scientists call this ***ballooning*** on silk thread.

Unlike silkworms, spongy moth caterpillars aren't picky eaters. Although they prefer oak leaves, they eat many kinds of leaves.

Spongy Moths are native to southern Europe, northern Africa, Japan, and parts of Asia.

In Étienne's time, there were no spongy moths in North America. To find out if his idea to breed them with silkworms would work, he imported some eggs.

When they arrived, he put them in a box on a windowsill in his workshop.

But then...

...a storm hit!

Lightning flashed.

Thunder boomed.

Wind howled.

Rain poured.

Guess what happened to the box of eggs Étienne put by an open window in his workshop!

The storm's wind scattered the eggs
in the woods behind Étienne's house.

The following spring, the eggs hatched.

Those caterpillars became moths that laid eggs.

The next spring, more caterpillars hatched.

These became moths, which laid more eggs.

Over again,

and again,

and again.

Since then, spongy moths have been slowly spreading across the United States and Canada. They are currently found in over 20 states and parts of Canada.

They are classified as an ***invasive species.***

What was considered a possible solution to a problem in the 1800s, has created a whole new set of problems and challenges today.

The red shows the spread of spongy moths across the United States and Canada from Massachusetts beginning in 1900.

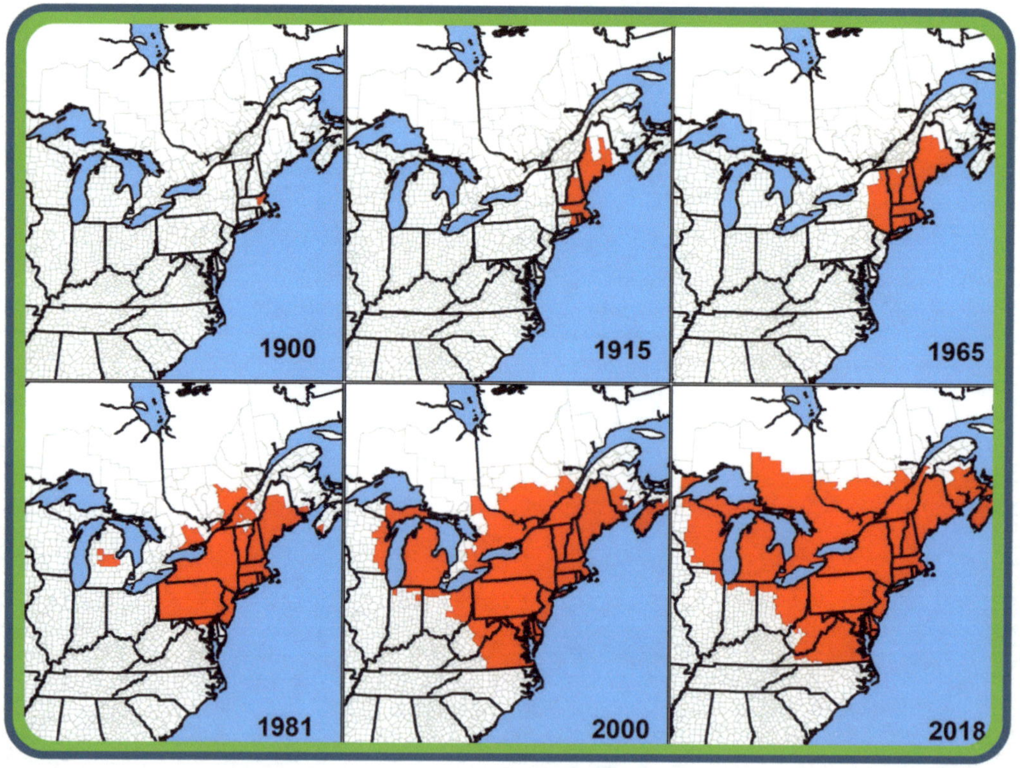

Most years, we hardly notice them. Numerous ***predators*** keep the spongy moth population low. As a result, there aren't enough to cause problems.

But some years, when certain conditions occur, the predator population declines.

Without enough predators, an *outbreak* can occur.

Yellow-billed and Black-billed Cuckoos, Blue Jays, Baltimore Orioles and Rufous-sided Towhees eat spongy moths while Chickadees feed on the egg masses.

Spongy moth pupae are a favorite food of white-footed mice.

When the caterpillars hatch in early spring, they are HUNGRY!

They spend their days eating LOTS of leaves.

Small, uneaten pieces of leaves fall like confetti, covering the ground.

During outbreak years, the ***ferocious*** appetites of millions of caterpillars ***defoliate*** many trees. This makes trees ***vulnerable*** to disease and other pests.

Two views of the same canopy, one week apart in May of 2024. *What do you notice?*

All this eating makes the caterpillars grow very quickly in just a few weeks.

Newly hatched caterpillars are about 0.25 inches. They are so light and furry, they can be carried long distances by the wind.

They also make LOTS of poop!

Small balls of poop, called frass, shower down from the trees. It sounds like light rain as it hits leaves, branches, and the ground.

Frass covers everything! Sidewalks and roads can get really slippery to walk, bike, or drive on.

Yuck!

The hood of this truck is covered with both frass and greenfall which are pieces of leaves cut by caterpillars that fall to the ground.

Do you notice the car keys near the front of the hood? They are there to give you a perspective of size.

After almost 3 weeks of non-stop eating, the caterpillars climb down the trees. They are now fully grown.

They are looking for a protected location for their ***pupal*** phase. They prefer a tree's bark crevice. But you might also find pupae on furniture, under building eaves, or in a car's bumper or tire.

They can be seen everywhere!

Eggs laid on a moving vehicle can be carried to other areas. This helps spongy moths to spread.

Then, they shed their skin.

This new skin becomes a hard, dark-brown shell. Inside this shell, they ***metamorphize*** and transform into adult moths.

The adults emerge in 12–16 days. They only live a few days and have one job: to mate! Only males fly. Although females have wings, they don't fly. To attract males, females send out special scents called ***pheromones***.

The air becomes filled with excited males. They fly in wild, zigzag patterns as they look for females.

Male

Female

Watch out!
They can fly into your face, hair, car, and house!

After mating, the females lay eggs.

Each egg mass can contain up to 1,000 tiny eggs. They can be found everywhere!

The eggs will hatch the following spring.

OH NO!

Not another summer of creepy-crawly caterpillars, raining poop, and moths doing face-plants on people and car windshields!

MAYBE NOT!

After a few years, the predator population begins to recover. With more predators, the spongy moths population begins to drop. Scientists call this a population ***boom-and-bust cycle.***

This cycle is complex and ***interconnected*** with many things. Oak trees, acorn production, and predators such as white-footed mice, play a role. So do some insects and parasites.

Two diseases also play an important role in the population collapse. A viral disease named **NPV** and a fungal disease named **Entomophaga maimaiga**. Weather plays a role in the effectiveness of the fungal disease.

Spongy moths can also cause their own ***demise***. As they defoliate the trees, less food becomes available. The lack of food causes many to die of starvation.

In this boom-and-bust cycle, the time between ***outbreaks*** is about every 10 years. It has also been as long as 30 years.

Dr. Clive Jones

is a scientist at the **Cary Institute of Ecosystem Studies.** This 2,000-acre research campus is in Millbrook, New York.

For 30 years, he has observed ***patterns***, analyzed data, and conducted experiments to learn about spongy moth outbreaks and why they collapse.

More research is needed to help us better understand how this complex web of interactions works. We also need to learn more about how this web influences the spongy moth, and how the spongy moth influences it.

Perhaps you will become a scientist who will help add to our knowledge of how our beautiful and complex natural world functions.

During a spongy moth outbreak, it might be tempting to blame Étienne Trouvelot. Especially since his idea of breeding a better silkworm wasn't successful.

He only wanted to help his friends and neighbors.

He had no idea his desire to solve a problem in his time would create a new one. One that would continue to affect people over 150 years later . . .

...because of a storm

Unintended Consequences

All actions have a result. Sometimes the result is planned and expected. When it is not, as in this story, the result is called an **unintended consequence**. Some unintended consequences, such as the spongy moth problem, have a negative impact. However, some have had positive results.

From Radar to Oven

Percy Spencer wasn't looking for a different way to cook food! He was trying to improve radar systems for the U.S. military. One day, while standing in front of an active radar set, he noticed the candy bar in his pocket had melted. To find out why, he set up experiments using popcorn and eggs. Over time, these experiments led to the invention of the microwave ovens found in today's kitchens.

Play-Doh Didn't Start as a Toy!

In the 1930s, brothers Noah and Cleo McVicker were trying to solve a problem: how to remove soot from the wallpaper without soap and water.

At that time, most families heated their homes with coal stoves. When burned, coal creates soot. This sticky, dirty substance coated everything, including walls and ceilings. Most walls were covered with wallpaper instead of paint. When washed with water, the wallpaper peeled off. The brothers solved this problem by creating a cleaner that didn't use water. It was made of soft clay mixed with detergent.

Over time, oil and gas furnaces replaced coal stoves. This ended the soot problem and the need for the clay cleaner. But there was still a lot of clay left over! In 1954, a nursery school teacher had an idea! She used the clay to make Christmas ornaments. This gave the brothers an idea. They removed the detergent, added color, and an almond scent. They renamed it Play-Doh and marketed it as a children's toy! The rest is history!

Super Glue
Was Considered a Failure Twice!

During World War I, engineer Harry Coover Jr. was trying to improve gun sights for the army. One formula he created was too sticky and didn't work.

In 1951, he was asked to develop a heat-resistant coating for jet planes. Remembering the sticky formula, he tried it. Too sticky again!

But the formula's stickiness got him thinking. Could it be used to stick things together? He experimented with different ideas for several years. Eventually, he introduced his sticky formula to the world as Super Glue.

The Failure that Led to Post-it Notes!

Spencer Silver was an engineer at 3M. His company was trying to develop super-strong adhesives for building airplanes. One adhesive he created was too weak. Although considered a failure, it had an interesting quality. It stayed sticky and could be reused.

A few years later, Art Fry, another 3M engineer, was singing in his church choir. His page markers were always falling out of his hymnbook. Remembering the weak, reusable adhesive, he had an idea. What if he put some on his page markers to hold them in place? Then, he could remove them when he finished.

The two engineers worked with the "failed" adhesive. They looked for ways to take advantage of the glue's weak, reusable quality. Their work led to the development of Post-it Notes. Today, Post-it Notes are one of the bestselling office supply products in the world.

More About Silk

Many insects produce silk. However, silk fabric is made from the Bombyx mori caterpillar. To make their cocoons, they produce silk in glands in their heads.

A silkworm eats about 30,000 times its weight in mulberry leaves.	It takes 3,000 cocoons to produce 1 pound of silk.
Each cocoon is made from a single thread that can be 2,000 - 3,000 feet long !	It takes 110 cocoons to make enough silk for a necktie, 630 for a blouse, and almost 5,000 for a dress.

Other insects that make silk include bees, wasps, and ants, silverfish, mayflies, thrips, leafhoppers, beetles, lacewings, fleas, flies, midges and arthropods such as spiders.

Who Figured Out How to Make Silk?

The Chinese began making silk over 4,500 years ago. No one knows who first unwound a silkworm cocoon and wove it into fabric. One legend tells us it was a Chinese empress. One day, as she sat under a mulberry tree sipping tea, a silkworm cocoon dropped into her cup. Captivated by the beautiful, shiny cocoon, she studied it. When she discovered it was made of thread, she began to unwind it. Over time, she realized the thread's strength and wove it into a fabric.

The fabric was so magnificent that only the emperor's family could wear it. For over 2,000 years, the Chinese kept the silk-making process a secret.

In time, Chinese traders began selling silk in Europe. Silk clothes became a symbol of wealth. The silk trade became so important that the trade route became known as the Silk Road. However, the silk-making process remained a secret. Anyone caught sneaking eggs, cocoons, or mulberry seeds out of China was put to death.

In the 5th century, two European monks visited China. While there, they secretly learned how to make silk. Before traveling home, they hid eggs and mulberry seeds in their walking canes! They managed to get out of China without getting caught. Back home in Europe, they shared their knowledge with others. Over time, Europeans learned how to raise silkworms and produce their own silk.

Silk in America

In 1609, King James I of England wanted to produce silk in his country. But it was too cold in England to grow mulberry trees. The king knew it was warmer in his Virginia colony. So he created a new law. It ordered tobacco farmers in the colony to plant mulberry trees and raise silkworms. But the colonists refused! They could make more money growing tobacco.

Many years later, after the colonies gained their independence from England, many families in New England began making silk at home. But disease, weather, and poor-quality fabric created problems. After the invention of the power loom in the 1870s, silk production moved to factories. Silk made on the power loom was smoother and better quality.

Thanks a Lot, Étienne!

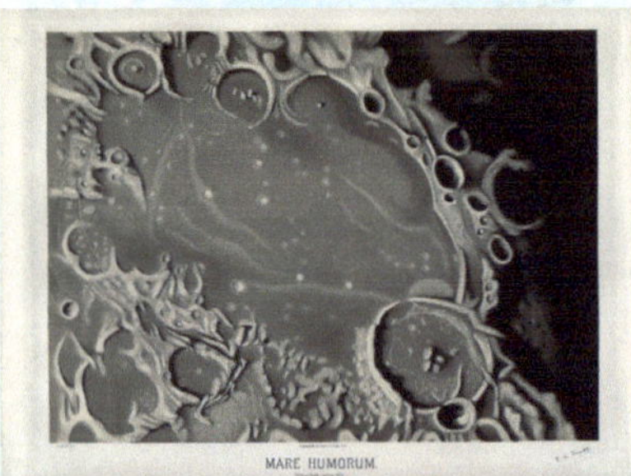

Étienne Léopold Trouvelot gave us more than the spongy moth!

He was also an astronomer and a skilled illustrator. In 1870, he saw a phenomenon known as the aurora borealis, also known as the northern lights. He was so amazed; he began illustrating them.

A few years later, the director of the Harvard College Observatory saw those illustrations. Impressed, he invited Étienne to join the staff.

Étienne made other contributions to science. In 1875, he discovered "veiled spots" on the sun. This helped us gain a better understanding of the sun. He created almost 7,000 space illustrations. He also wrote nearly 50 scientific papers.

Hey Étienne! Why didn't you use a screen in your window?

It is likely that a screen would have prevented Étienne's spongy moth eggs from being blown out the window. However, screens were not commonly used in homes until the late 1800s. Air conditioning wasn't developed until the early 1900s. Therefore, people left windows open to allow air to flow through the house. The development of air conditioning is also an example of an unintended consequence.

Étienne Léopold Trouvelot's home in Medford, MA, From "The Gypsy Moth," by E.H. Forbush and C.H. Fernald, 1896

That's a Weird Name!

How the Spongy Moth Got Its Name

Until recently, spongy moths were called gypsy moths. In Étienne Léopold Trouvelot's time, the spelling was "gipsy." However, in recent times, many people considered the term "gypsy" an insult.

In 2021, the Entomological Society of America (ESA) decided to change it. They brought together a team of scientists, forestry experts, and scholars. Their research showed that "sponge" was part of this insect's name in many countries. Why? Because the moth's light brown, fuzzy egg mass looks like a sponge. This led them to adopt spongy moths as their new common name.

Invasive Species

What's the Big Deal?

An invasive species is a plant, animal, or insect not native to an ecosystem. In their native ecosystem, they are not a problem. There, they have predators to keep their population in balance. But when taken to another ecosystem, they can become a problem. There may be no predators to keep their population under control. They often compete with native species for food, water, and shelter.

Spongy moths are one of over 4,300 invasive species in the United States. Invasive species are the second-biggest threat to the health of ecosystems worldwide.

The Spongy Moth's Boom-And-Bust Cycle

The spongy moth's boom-and-bust cycle takes place in woodland ecosystems. It is a complex web of many organisms and interactions.

Oak Leaves
Spongy moths eat many types of leaves. But oak leaves are their favorite.

Acorns
Acorns are the nuts of the oak tree. They are rich in nutrients and are a key food source for white-footed mice and other rodents.

Weather
Temperature and precipitation amounts appear to affect the growth and quantity of acorns.

Severe winter temperatures can kill egg masses. Warmer winter temperatures enable them to survive. Weather also affects one of the spongy moth's predators.

Predators

Birds
Not all birds like to eat the large, hairy caterpillars. Fortunately, there are a few! Yellow-billed and Black-billed Cuckoos, Blue Jays, Orioles, and Rufous-sided Towhees all eat them. Black-capped Chickadees eat spongy moth egg masses.

Mammals
Shrews, mice, voles, chipmunks, skunks, raccoons, and squirrels all eat spongy moth caterpillars, and pupae.

Insects
Stink bugs, ants, and spiders eat spongy moth caterpillars and pupae.

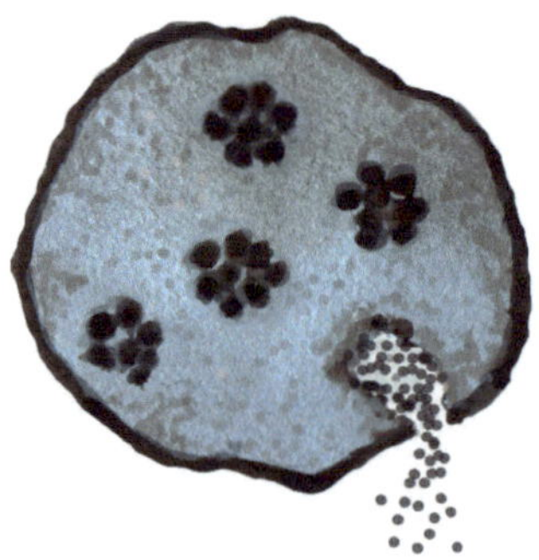

NPV
NPV is a virus. It is always present in spongy moth populations. The female moth can pass it on to her offspring. During an outbreak, food and space become scarce. This stresses and weakens the caterpillars making them unable to protect themselves from the virus. Caterpillars killed by NPV hang in an upside-down V shape. These bodies crumble, spreading the virus.

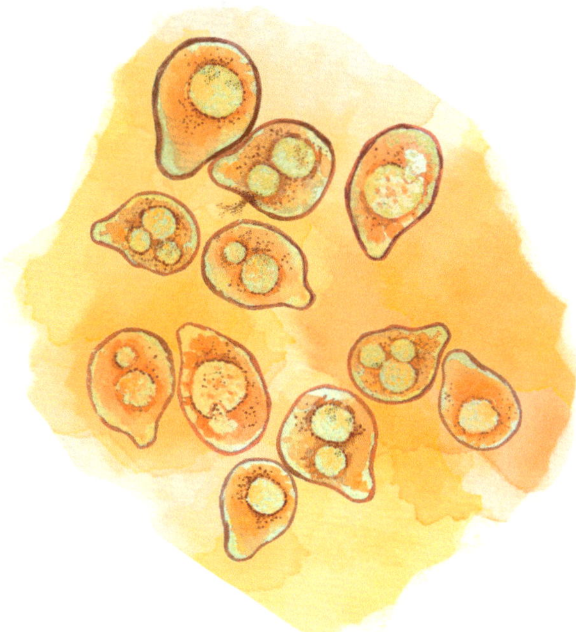

Entomophaga maimaiga

Entomophaga means "insect eater" in Latin. *Maimaiga* is the Japanese name for the spongy moth. Together, "an eater of spongy moths."

Maimaiga (my-MY-yigh-ga) for short! The spores of this fungus overwinter in the soil. A cool, wet spring creates the perfect conditions for the spores to spread and infect young caterpillars. However, a hot, dry spring will reduce the spores' spread.

Caterpillars killed by this fungus hang upside down. Their stiff, brittle bodies fall to the ground. There, they break apart, releasing the spores. The wind carries the spores and infects more caterpillars.

During an outbreak, soil infected with spores is collected. Then it is taken to other areas and released, helping the spores to spread.

Scientists are still studying and learning about the spongy moth cycle. There may be other organisms and interactions that play a role.

Scientists do know that predators play a key role in this cycle. Predators do well with plenty of food, along with the right conditions and weather. That's good! Lots of predators eat lots of caterpillars and pupae. This leds to fewer adult moths laying eggs.

But different weather, conditions, and less food availability can cause the predator population to decline. That's not so good!

Fewer predators mean more caterpillars, pupae, and adult moths laying eggs. This could lead to a population boom and a possible outbreak the following year.

The flowchart on the next page illustrates the spongy moth cycle.

Track each pathway, noting the conditions to discover which might lead to an outbreak.

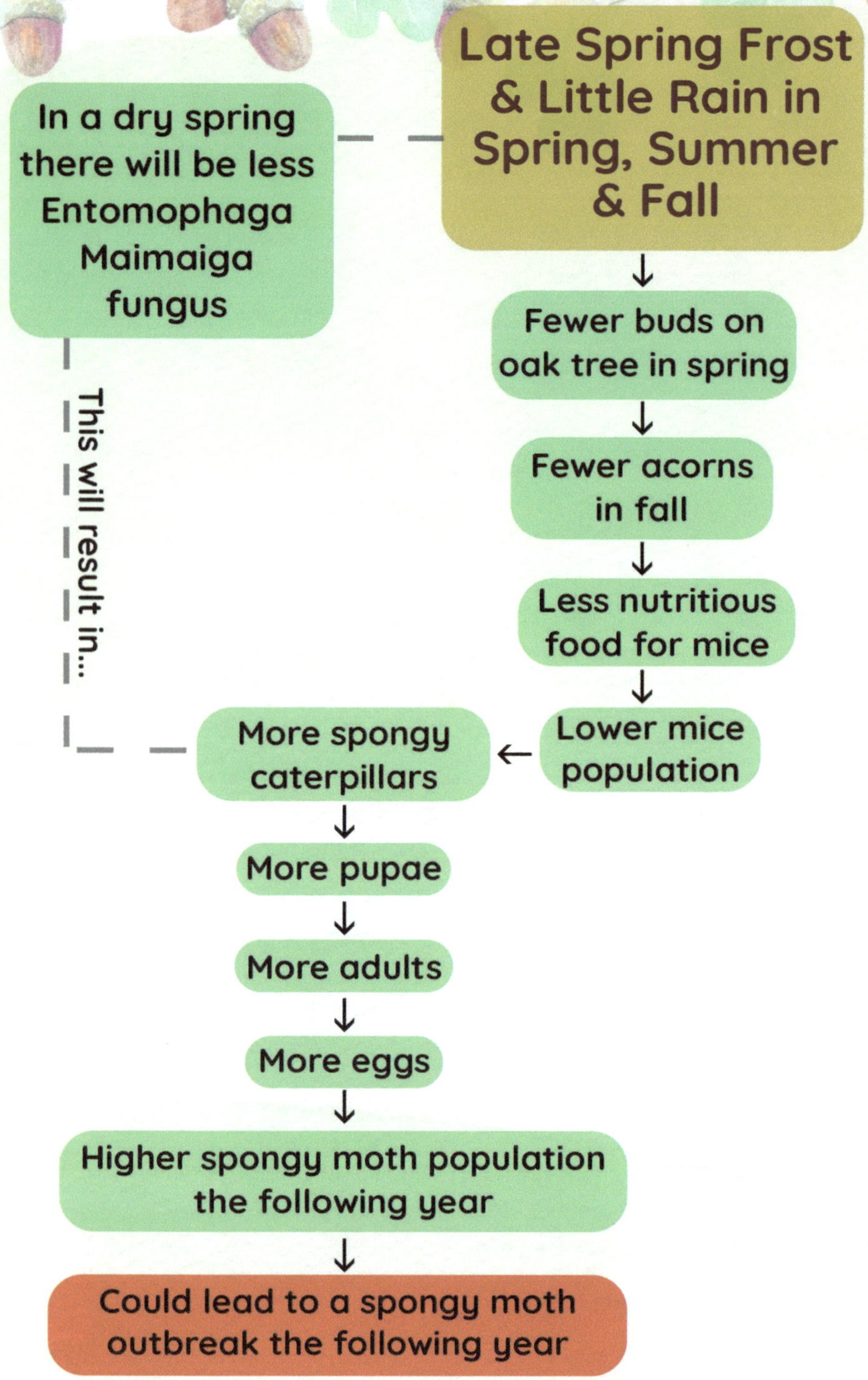
Late Spring Frost & Little Rain in Spring, Summer & Fall
In a dry spring there will be less Entomophaga Maimaiga fungus
This will result in...
Fewer buds on oak tree in spring
Fewer acorns in fall
Less nutritious food for mice
Lower mice population
More spongy caterpillars
More pupae
More adults
More eggs
Higher spongy moth population the following year
Could lead to a spongy moth outbreak the following year

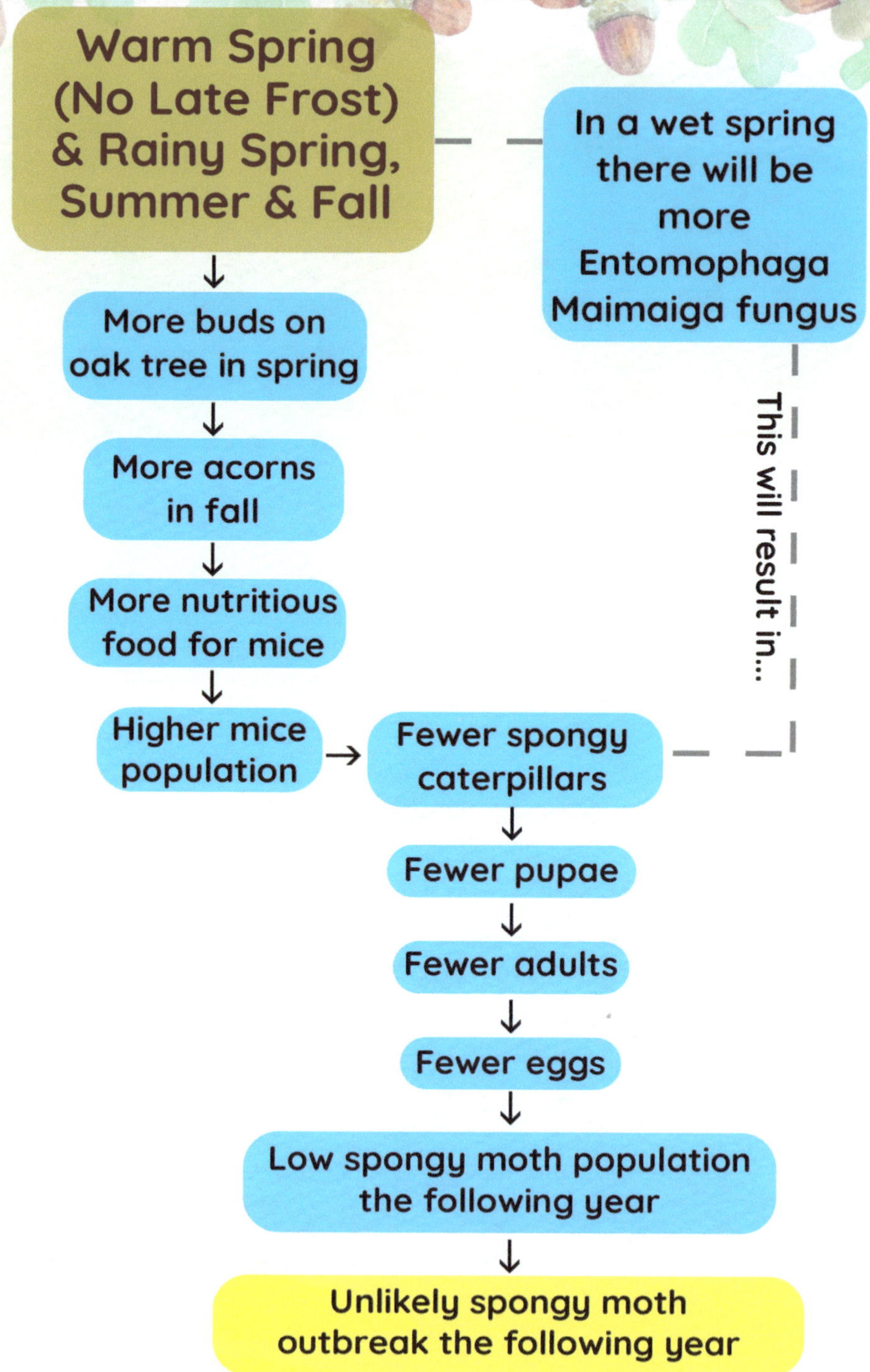
Warm Spring (No Late Frost) & Rainy Spring, Summer & Fall
In a wet spring there will be more Entomophaga Maimaiga fungus
This will result in...
More buds on oak tree in spring
More acorns in fall
More nutritious food for mice
Higher mice population
Fewer spongy caterpillars
Fewer pupae
Fewer adults
Fewer eggs
Low spongy moth population the following year
Unlikely spongy moth outbreak the following year

Fighting Back

Spongy moths are here to stay. We can't stop them. But there are several ways to slow their spread. These methods were developed using what scientists and biologists have learned about spongy moths.

Burlap Traps

Burlap traps were developed from what we know about how the caterpillars move. Traps are made of a of 12 - 18" wide burlap strips, long enough to wrap around a tree. A string is tied around the center of the fabric. The top of the fabric is folded over the string. Caterpillars climb the tree and get stuck inside the burlap.

Sticky Bands

Sticky bands are wrapped around the tree with the sticky side out. As caterpillars climb the tree, they get stuck on the tape. Unable to move, they die.

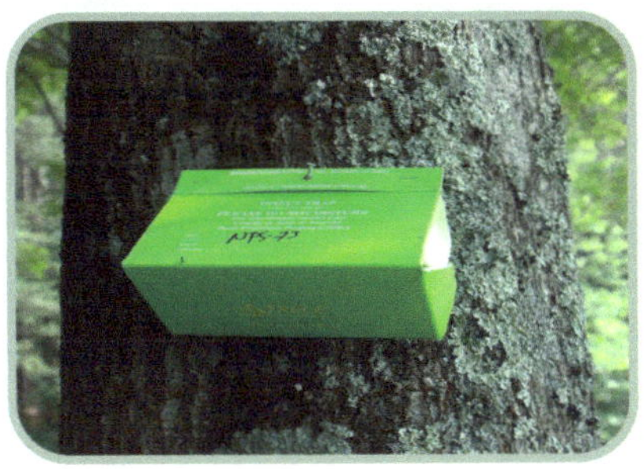

Pheromone Traps

Pheromones are chemicals that help insects communicate. When ready to mate, female spongy moths release them to attract males. Scientists have learned how to reproduce pheromones. These man-made pheromones are put in trap boxes. Male moths are attracted to them and get caught in the box.

Squishing

You don't need research to know that squishing and scraping work! Squishing is yucky! It also takes time to squish all those caterpillars, moths, and egg masses. Although it reduces the population, it won't stop them from spreading because there are too many.

Bacillus Thuringiensis (BtK)

This is the most common way to treat spongy moths. It is a bacterium that interferes with the spongy moth caterpillars' digestive system and eventually kills them. BtK is found naturally in soil and used to make a spray. It doesn't affect other animals and degrades when exposed to sunlight.

Story Resources

- *Cary Institute Spongy Moth Chronicles* **https://www.caryinstitute.org/topics/spongy-moth**
- *Natural Enemies of Gypsy Moth: The Good Guys!* Michigan State University Extension Bulletin E-2700, April 1999
- *Chain Reactions Linking Acorns to Gypsy Moth Outbreaks and Lyme Disease Risk* by Clive G. Jones, Richard S. Ostfeld, Michele P. Richard, Eric M. Schauber, and Jerry O. Wolff, Science, Feb 13, 1998, Vol. 279, No. 5353, pp. 1023–1026, DOI: 10.1126/science.279.5353.1023 **https://www.science.org/doi/10.1126/science.279.5353.1023**
- *A Virus and a Fungal Disease Cause Spongy Moth Outbreaks to Collapse*, April 27, 2022, Deborah McCullough **https://www.canr.msu.edu/resources/a-virus-and-a-fungal-disease-cause-spongy-moth-outbreaks-to-collapse**
- *The Gipsy Moth and Its Introduction into America,* Nature Journal, Vol. 60, No. 1543, May 25, 1899
- Connecticut's Mulberry Craze, Bob Wyss for Connecticut Explored, **https://connecticuthistory.org/connecticuts-mulberry-craze/**

Other Reader Resources

- Kentucky Citizen Science Invasive Species Program **https://storymaps.arcgis.com/stories/39ac8e6315a241ff9fcc26bdda444f51**
- Tree-SMART Trade **https://www.caryinstitute.org/science/tree-smart-trade**
- Help Stop the Spread of the Spongy Moth! **https://plantheroes.org/educator_resources/spongy-moth-activity-8-10/**

Photo Credits

Page 3	Silk Clothing	AI Generate Images
Page 9	Étienne Léopold Trouvelot	Public domain, via Wikimedia Commons https://commons.wikimedia.org/wiki/File:%C3%89tienne_L%C3%A9opold_Trouvelot.jpg
Page 10	Eggs, Larva	Photo by: Bill McNee/Wisconsin DNR - Used by permission
Page 10	Adult	USDA APHIS Plant Protection and Quarantine , USDA APHIS PPQ, Bugwood.org - Used by permission
Page 10	Pupa	Photo by: M. Zubrik, NFC from Forestry Images - Used by permission
Page 11	Ballooning Caterpillar	Photo by: Rick Koval
Page 16	Map	Image used by permission of US Forest Service Source: *Slow the Spread: a national program to manage the gypsy moth.* Chapter 2. pg. 21 *Population Biology of Gypsy Moth Spread*; Andrew M. Liebhold, Alexei A. Sharov, and Patrick C. Tobin https://research.fs.usda.gov/treesearch/12583
Page 17	Blue Jay	https://commons.wikimedia.org/wiki/File:Blue_jay_in_Central_Park_(16465).jpg Rhododendrites, CC BY-SA 4.0 <https://creativecommons.org/licenses/by-sa/4.0>, via Wikimedia Commons

Page 17	White-footed mice	Kelly Oggenfuss. © Cary Institute of Ecosystem Studies used with permission. https://www.caryinstitute.org/sites/default/files/public/2024-06/smc_6_19_24_photo_3.jpg
Page 18	Caterpillars on oak leaf	Photo by: Bill McNee/Wisconsin DNR - Used by permission
Page 18	Defoliated tree canopy	Michael Fargione. © Cary Institute of Ecosystem Studies used with permission. https://www.caryinstitute.org/sites/default/files/public/2024-06/smc_.may_30_2.jpg
Page 20	Frass on car hood	Michael Fargione. © Cary Institute of Ecosystem Studies used with permission. https://www.caryinstitute.org/sites/default/files/public/2024-05/smc_5_21_car_4.jpg
Page 20	Frass on picnic table	Photo by: John Ghent, John Ghent, Bugwood.org - Used by permission
Page 21	Caterpillars on tree	Photo by: Adrian Bryan-Brown © 2023
Page 21	Caterpillars on tire	Photo by: Michael LaMonica licensed through Shutterstock
Page 21	Caterpillars on plant pot	Photo by: Bill McNee/Wisconsin DNR - Used by permission
Page 21	Caterpillars on building	Photo by: Akchamczuk licensed through iStock by Getty Images
Page 22	Male & Female	Photo by: Jeffrey W. Lotz, Florida Department of Agriculture and Consumer Services, Bugwood.org used by permission
Page 22	Pupa	Photo by: Gyorgy Csoka, Hungary Forest Research Institute, Bugwood.org- Used by permission
Page 23	Females laying eggs	Photo by: Milan Pernek, Bugwood.org - Used by permission
Page 23	Closeup of eggs	Photo by: Bill McNee/Wisconsin DNR - Used by permission
Page 25	Dr. Clive Jones	© Clive G. Jones used with permission. https://www.caryinstitute.org/sites/default/files/public/styles/scientist_photo_large/public/2024-06/bio_jonesclive.jpg
Page 30	Microwave Oven	Shutterstock Image by Inna Kharlamova
Page 30	Play-dough	Hasbro Used by Permission
Page 31	Super Glue	Author: Omegatron, Licence: Creative Commons Attribution-Share Alike 3.0 Unported, CC BY-SA 3.0, https://commons.wikimedia.org/w/index.php?curid=640168
Page 31	Post-Its	Shutterstock Image by 32 pixels
Page 33	King James	Public domain, via Wikimedia Commons https://en.wikipedia.org/wiki/James_VI_and_I
Page 34	"Total Eclipse of the Sun"	Étienne Léopold Trouvelot, Public domain, via Wikimedia Commons, https://commons.wikimedia.org/wiki/File:Trouvelot_-_The_Total_Eclipse_of_the_Sun.jpg
Page 34	"The planet Saturn"	Étienne Léopold Trouvelot, Public domain, via Wikimedia Commons, https://commons.wikimedia.org/wiki/File:Trouvelot_-_The_planet_Saturn_-_1874.jpg
Page 34	"Mare Humorum"	Étienne Léopold Trouvelot, Public domain, via Wikimedia Commons, https://commons.wikimedia.org/wiki/File:Trouvelot_-_Mare_Humorum_-_1875.jpg
Page 34	"Aurora Borealis"	Étienne Léopold Trouvelot, Public domain, via Wikimedia Commons, https://commons.wikimedia.org/wiki/File:%C3%89tienne_L%C3%A9opold_Trouvelot,_%E2%80%9CAurora_Borealis,_from_The_Trouvelot_Astronomical_Drawings%E2%80%9D_2006_48_4v1.jpg
Page 34	Etienne Léopold Trouvelot home	"The Gypsy Moth," by E.H. Forbush and C.H. Fernald, 1896 (sourced from: *https://www.thoughtco.com/how-the-gypsy-moth-came-to-america-1968402)*

Dorna Schroeter was coordinator of the PNW BOCES Center for Environmental Education for 38 years. The program served some 40,000 students each year with a wide diversity of programs. Her favorite was the marine ecology program she ran several times a year in the Florida Keys. Since retiring, she has authored *We Are One,* and *The Story of Velcro and The Story of Eco-Machines,* two books about biomimicry in the series *How an Idea From Nature Changed Our World.* She has also authored several articles in scientific and professional publications. When she isn't writing stories, teaching biomimicry, mentoring youth, or gardening, Dorna can be found biking the country roads around her home and on multi-day trips on the many rail trails around the country.

Illustrator Natalia Chervochkova, has told children many stories with the help of illustrations. She works in the field of book graphics and marketing projects for children for over 10 years. She loves to spend time in her rose garden and taking care of plants. Natalia believes that the most complex things can be explained to children in the simple language of pictures and humor.

Thank You!

A conversation with friends inspired this story. We were sitting in our village courtyard on a beautiful June day in 2024, watching caterpillars ballooning, listening to frass fall, and wondering how bad the outbreak would be.

Writing and publishing a story takes a village. I'm blessed to have a wonderful group of friends and colleagues who support me. Mary Burns, Diane Madaio, Patty Dannerman, Robin Shornstein, Mary Ellis, Mary Burns, Lily Ross, Diane Duffus, Joanne Gelb, Rachel Arbor, members of Likes' Publishing critique group, especially SaDonna Heathman, who always asked just the right questions and offered the right suggestions to move me forward.

Special thanks to:

- **Dr. Clive Jones and his team at the Cary Institute of Ecosystem Studies in Millbrook, NY for their patient and careful scientific review and input.**
- **The many photographers who enthusiastically and generously gave me permission to use their photos.**

Vocabulary

Ballooning: Young caterpillars use this method to move from one tree to another. They produce a silk thread and hang from it. With light body weight and long hair, the wind can carry them for miles to new locations.

Boom-and-bust cycle: Periods of time when population numbers are high, followed by periods of relatively low numbers.

Defoliates: The removal of a tree's leaves.

Demise: To die.

Entomologist: a scientist who studies insects, their lives, and how they interact with humans and the environment.

Exoskeleton: An external supporting structure of a living creature, such as the shell of a crustacean.

Ferocious: Fierce, savage.

Interconnected: The relationship of one to others.

Invasive species: A plant, animal, or insect that isn't native to an ecosystem and can disrupt the balance of its new environment. It can be brought into a new ecosystem in several ways, including weather and human actions.

Iridescent: Shiny, gleaming, shimmery.

Metamorphize: The process most insects go through to become adults.

Outbreak: A sudden, intense increase of something, such as a disease or animal population.

Patterns: Things that are repeated in a predictable way.

Pheromones: Chemicals that insects produce to communicate. Some are detected by smell, and others by taste.

Plague: A deadly disease that spreads quickly; a sudden invasion of harmful insects.

Predator: An animal that hunts other animals for food.

Pupal: Last stage of an insect's life cycle before becoming an adult. While in this stage, they do not eat or move.

Vulnerable: Able to be hurt or injured.

www.ingramcontent.com/pod-product-compliance
Lightning Source LLC
Chambersburg PA
CBHW041317210926
51847CB00004B/49
9781966997009